Journal of Approximation Theory and Applied Mathematics

2014 Vol. 4

Contents

Approximation Error by Using a Finite Number of Base Coefficients for Special Types of Wavelets

Solving Fredholm Integral Equations with Application of the Four Chebyshev Polynomials

Fourier Properties of Approximations with Functions on a Compact Interval using Daubechies Wavelets

Herstellung und Verlag:
BoD – Books on Demand, Norderstedt
ISBN: 978-3-7347-4403-7

Approximation Error by Using a Finite Number of Base Coefficients for Special Types of Wavelets

M. Schuchmann and M. Rasguljajew from the Darmstadt University of Applied Sciences

Abstract

If an approximation y_j of y is determined by an orthogonal projection from y on V_j then in practical cases only a finite number of bases coefficients can be used. Here we investigate the relationship of the approximation error, resulting from the use of a finite number of basic coefficients, depending on the number of basic elements. We consider wavelets with compact support, such as Daubechies wavelets and the Shannon wavelet for different types of functions.

Introduction of the MSA

In the wavelet theory a scaling function ϕ is used, which belongs to a MSA (multi scale analysis). From the MSA we know, that we can construct an orthonormal basis of a closed subspace V_j, where V_j belongs to a the sequence of subspaces with the following property:

$$\ldots \subset V_{-1} \subset V_0 \subset V_1 \subset \ldots \subset L^2(R),$$

$\{\phi_{j,k}(t)\}_{k \in Z}$ is an orthonormal basis of V_j with $\phi_{j,k}(t) = 2^{j/2}\phi(2^j t - k)$.

We use the approximation function

$$y_j(t) := \sum_{k=-k_{max}}^{k_{max}} c_k \cdot \phi_{j,k}(t) \quad .$$

With the scaling function ϕ of the MSA we get an orthonormal basis of V_j with $\phi_{j,k}(t) = 2^{j/2}\phi(2^j t - k)$. So we get the orthogonal projection from a $L^2(R)$ function f in V_j with

$$y_j(t) = \sum_k c_k \phi_{j,k}(t) \text{ with } c_k = \langle y, \phi_{j,k} \rangle = \int_{-\infty}^{\infty} y(t) \cdot \overline{\phi_{j,k}(t)} dt \quad .$$

Approximation Error Through a finite k_{max}

If we use a finite number k_{max} than we get an approximation error for y in V_j depending on the k_{max}:

$$e_{k_{max}} = \left\|\sum_k c_k \phi_{j,k} - \sum_{|k| \leq k_{max}} c_k \phi_{j,k}\right\| = \left\|\sum_{|k|>k_{max}} c_k \phi_{j,k}\right\| = \sqrt{\sum_{|k|>k_{max}} |c_k|^2}$$

Now we consider a father wavelet with compact support: *supp* $\phi = [0,a]$ with $a > 0$.

For Daubechies wavelets of order *r* we get $a = 2r - 1$.

Considering Functions with an exponential decay

If $|y(t)| \leq c \cdot e^{-d \cdot |t|}$ (with $d > 0$ and $c > 0$) than we get for $k > 0$ with $m = max \, |\phi(t)|$

$$|c_k| \leq c \cdot m \cdot 2^{j/2} \int_{k/2^j}^{(a+k)/2^j} e^{-d \cdot t} dt = \frac{c \cdot m \cdot 2^{j/2} \cdot e^{-d \cdot k \cdot 2^{-j}}}{d} \cdot (1 - e^{-d \cdot a \cdot 2^{-j}}) \ .$$

We assume that k_{max} is so big, that the *supp* $\phi_{j,-k_{max}} \subseteq R^-$. If we use the maximum $|c_k|$ (with an equality in the equation above), we get the following relation between the approximation error and k_{max}

$$ln(e_{k_{max}}) - ln(e_{k_{max}+1}) = d \cdot 2^{-j}$$

which is independent from *c, m* and *a*. Here we assume additionally that $a > k_{max}$ and *y* is analytic (see after next chapter).

For a bigger *j* we need a bigger k_{max}, by rising *j* by 1 we need the factor 2 for k_{max}.

Now we see the Graph of the points (k_{max}, $max \, ln(e_{k_{max}}^2)$) (where we set the c_k on their maximum) joined by a line for $k_{max} = 5, 6, ..., 20$ and different *j*. For $j = 1$ we get the slope $-2^{-1} = -0.5$, for $j = 2$ we get the slope $-2^{-2} = -0.25$ and for $j = 3$ we get the slope $-2^{-3} = -0.125$.

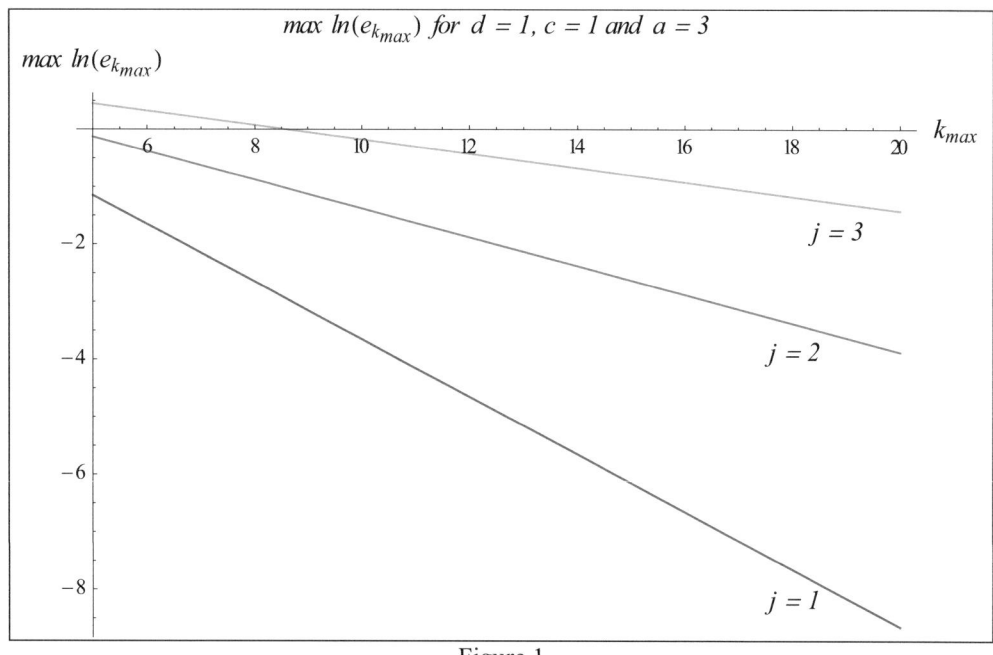

Figure 1

Considering Functions with a decay like $c \cdot (1 + t^2)^{-1}$

We assume that k_{max} is so big, that the $supp\ \phi_{j,-k_{max}} \subseteq R^-$. For an analytic function y with $|y(t)| \leq \dfrac{c}{1+t^2}$ we would get the following graphs if we (where we set the c_k on their maximum):

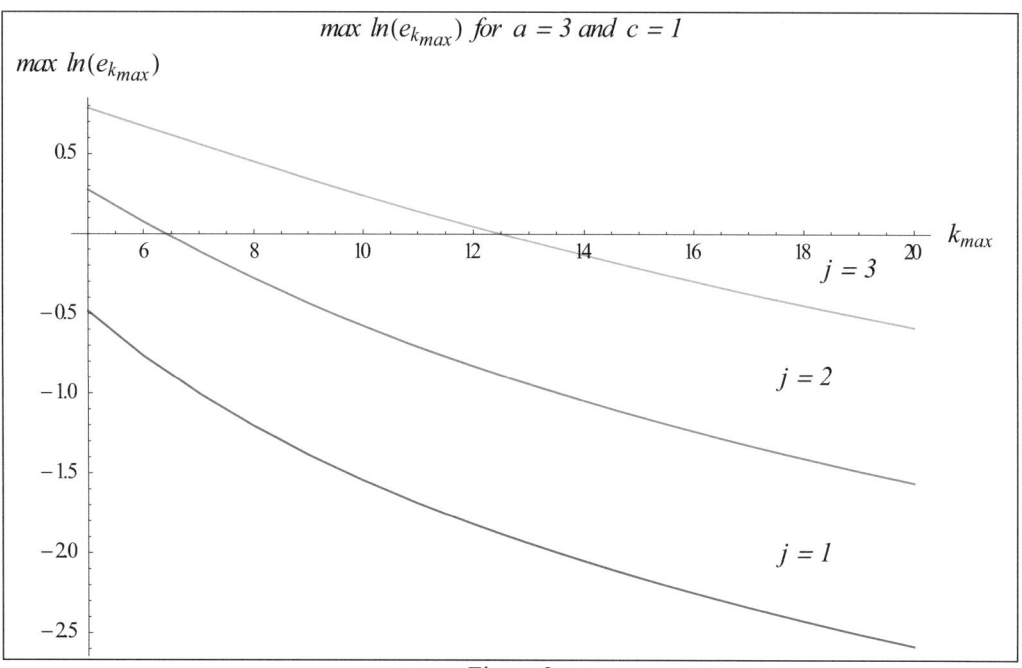

Figure 2

If we use the Shannon wavelet we get a similar graph for this type of function y with $c = 1$:

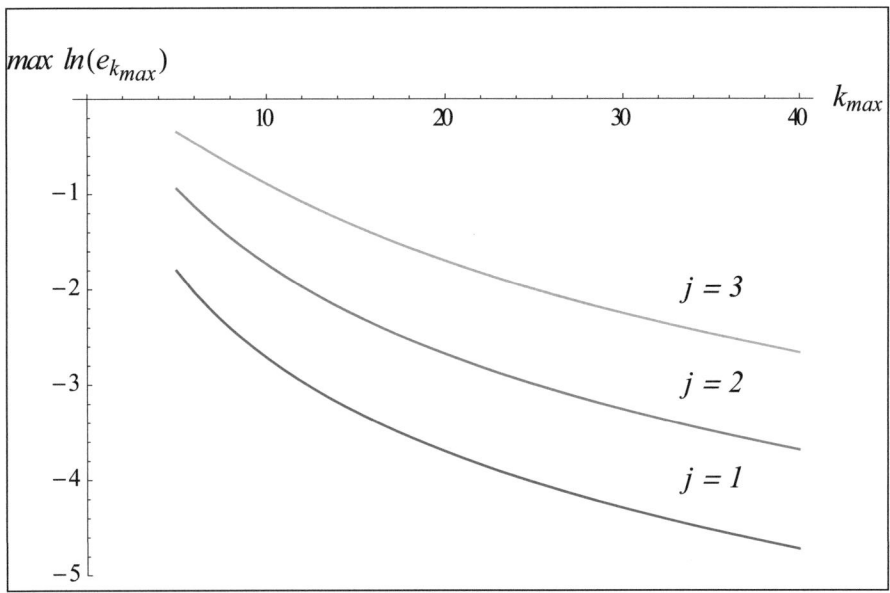

Figure 3

Error estimations in dependency of k_{max} under certain conditions for the decay behaviour can be found in [1].

General Approximation Error Through a finite k_{max} with the Shannon wavelet

If y is in V_j than the coefficients c_k are function values of y because of the Shannon theorem. Here we get $c_k = 2^{-j/2} \cdot y(2^{-j} \cdot k)$. So the decay behaviour of the coefficients depends on the decay behaviour of the function y.

If y would be not in V_j and j is big enough, so that y_j is a good approximation for y, then $c_k \approx \tilde{c}_k := 2^{-j/2} \cdot y(2^{-j} \cdot k)$. In that case the Fourier transform Y of y is for arguments $|\omega| > 2^j \cdot \pi$ nearly zero. The reason is:

$$\|y - y_j\|_{L^2} = \sqrt{\int_{-\infty}^{-2^j \pi} |Y(\omega)|^2 d\omega + \int_{2^j \pi}^{\infty} |Y(\omega)|^2 d\omega} \quad .$$

For a continues function y we get:

$$y(t) - y_j(t) = \frac{1}{\sqrt{2\pi}} \int_{-\infty}^{\infty} Y(\omega) e^{i\omega t} d\omega - \frac{1}{\sqrt{2\pi}} \int_{-2^j \pi}^{2^j \pi} Y(\omega) e^{i\omega t} d\omega$$

$$= \frac{1}{\sqrt{2\pi}} \int_{\{\omega | |\omega| > 2^j \cdot \pi\}} Y(\omega) e^{i\omega t} d\omega$$

So:

$$\left|\tilde{c}_k - c_k\right| = \left|2^{-j/2} \cdot y(2^{-j} k) - 2^{-j/2} \cdot y_j(2^{-j} k)\right| = \left|\frac{2^{-j/2}}{\sqrt{2\pi}} \int_{\{\omega | |\omega| > 2^j \cdot \pi\}} Y(\omega) e^{i \cdot 2^{-j} \cdot k \cdot \omega} d\omega\right|$$

Even for other Wavelets like Daubechies wavelets we get for a big enough j the approximation $c_k \approx \widetilde{c}_k$. If the first central moment of the scaling function ϕ is not zero we must shift the k.

Considering Functions with a decay like the Gauss function

Now we consider at first the Shannon wavelet and we assume that $|y(t)| \leq exp(-t^2)$.

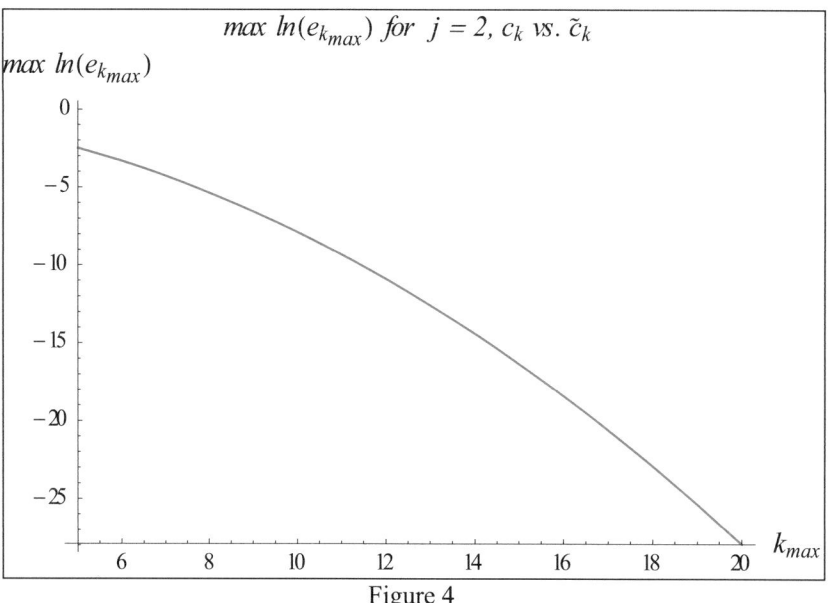

Figure 4

In the Figure above we see the maximum of $ln(e_{kmax})$. If we use the coefficients $\widetilde{c}_k = 2^{-j/2} \cdot y(2^{-j} \cdot k)$ then we see here graphically no difference.

With the Daubechies wavelet of order 4 we get the following graph for the maximum of $ln(e_{kmax})$:

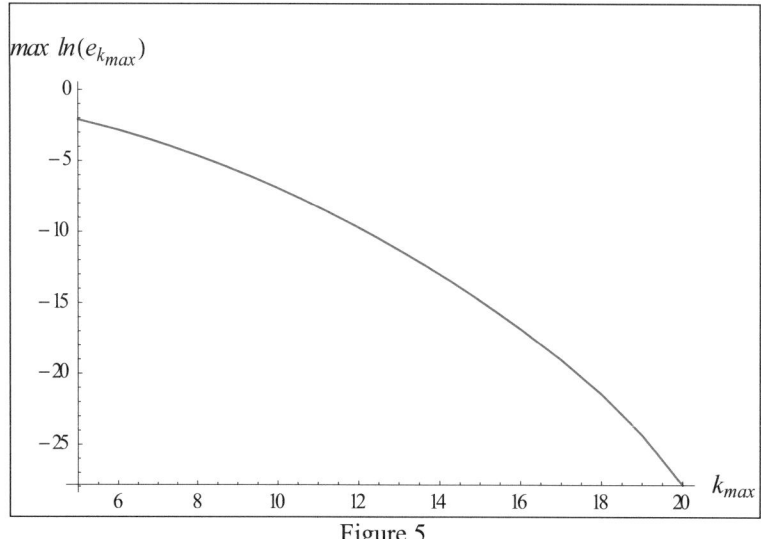

Figure 5

If we use the Shannon wavelet (we get analogous results with the Daubechies wavelet) and if we compare the approximation error for $y(t) = exp(-t^2)$ we see that from a certain k_{max} we have almost no improvement of the approximation. Here we consider the interval [-3, 3].

We see the graphs of $y - y_j$ with different k_{max}:

$j = 1$ and $k_{max} = 4$:

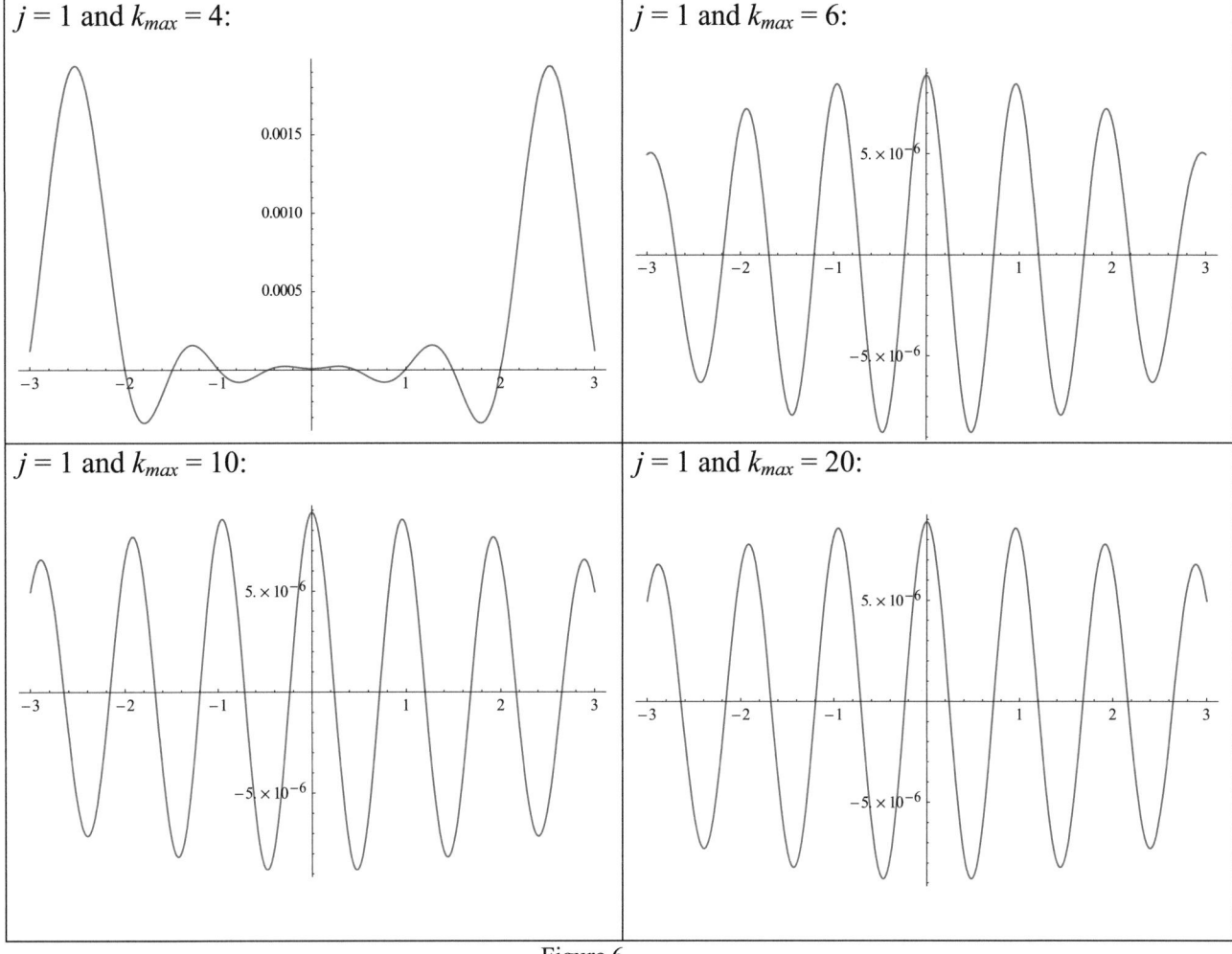

Figure 6

With bigger a j we need a bigger k_{max} to get an equally good approximation. If we increase j by 1 we would have to double k_{max}. From a certain k_{max} we get almost no improvement of the approximation if we take a look at the approximation error.

Here we see the a table with $||y - y_j||_{L^2[-3,3]}$ for different k_{max} where the integrals have been numerically evaluated (for $j = 1$ and $j = 2$):

k_{max}	$\|y - y_1\|_{L^2[-3,3]}$
4	0.00187818
5	0.000084134
6	0.0000132573
7	0.0000136874
8	0.0000138086
9	0.0000138871
10	0.0000139394
11	0.0000139756
12	0.0000140015
13	0.0000140205
14	0.0000140348
15	0.0000140457

k_{max}	$\|y - y_2\|_{L^2[-3,3]}$
10	0.000372283
11	0.0000576393
12	$3.02653 \cdot 10^{-6}$
13	$3.54982 \cdot 10^{-7}$
14	$4.29479 \cdot 10^{-8}$
15	$4.91575 \cdot 10^{-9}$
16	$5.16728 \cdot 10^{-10}$
17	$4.92355 \cdot 10^{-11}$
18	$4.18993 \cdot 10^{-12}$
19	$3.71437 \cdot 10^{-13}$
20	$1.27043 \cdot 10^{-13}$
21	$1.30018 \cdot 10^{-13}$
22	$1.297 \cdot 10^{-13}$
23	$1.29699 \cdot 10^{-13}$
24	$1.29699 \cdot 10^{-13}$
25	$1.29699 \cdot 10^{-13}$
26	$1.29699 \cdot 10^{-13}$
27	$1.29699 \cdot 10^{-13}$
28	$1.29699 \cdot 10^{-13}$
29	$1.29699 \cdot 10^{-13}$
30	$1.29699 \cdot 10^{-13}$
31	$1.29699 \cdot 10^{-13}$
32	$1.29699 \cdot 10^{-13}$

For the Daubechies wavelet of order 4 we get the following approximation error:

k_{max}	$\|y - y_1\|_{L^2[-3,3]}$
3	0.0750862
4	0.0175333
5	0.0129922
6	0.0129353
7	0.0129351

k_{max}	$\|y - y_2\|_{L^2[-3,3]}$
8	0.00977717
9	0.00343032
10	0.00138179
11	0.00100603
12	0.000973809
13	0.000972563

Approximation of non-continues functions of type $1_I \cdot y$

If y is not continues we have a bad decay behaviour (see [6] to [8]). If we consider for example a function of the type $1_I y$ (with $I = [a, b]$) and if we have jumps at the edges of I then the coefficients c_k have only the order $O(1/|k|)$. A function of that type could be used, if we need an approximation on a compact interval and if y is not quadratic integrabel on R. We now consider the Shannon wavelet in such a case:

If we use the Shannon wavelet and assume for $I = [-a, a]$, $a > 0$ and $k_0 := k_{max} - \lceil 2^j a \rceil$ where $\lceil x \rceil$ is the smallest integer n with $n \geq x$. $c_k = \langle 1_I y, \phi_{j,k} \rangle$. Additional we assume, that $k_{max} > \lceil 2^j \cdot a \rceil$. Then we get the following error estimation:

$$\left\| \sum_k c_k \phi_{j,k} - \sum_{|k| \leq k_{max}} c_k \phi_{j,k} \right\| \leq \frac{2^{1+j/2} c}{\pi} \sqrt{\zeta(2) - \sum_{k=1}^{k_0} \frac{1}{k^2}}$$

Here ζ is the Riemann ζ function and c depends on a and y.

The reason is:

$$|c_k| = \left| \int_I y(t) \cdot \phi_{j,k}(t) dt \right| \leq \int_I |y(t) \cdot \phi_{j,k}(t)| dt$$

$$\leq 2a \cdot \max_{t \in I} |y(t) \cdot \phi_{j,k}(t)| \leq \underbrace{2a \cdot \max_{t \in I} |y(t)|}_{:=c} \cdot \max_{t \in I} |\phi_{j,k}(t)|$$

Now we consider the last factor:

$$\max_{t \in I} |\phi_{j,k}(t)| = 2^{j/2} \cdot \max_{t \in I} \left| \frac{\sin(\pi(2^j t - k))}{\pi(2^j t - k)} \right| \leq \frac{2^{j/2}}{\pi} \max_{t \in I} \left| \frac{1}{2^j t - k} \right|$$

Because of symmetry we consider only $k > 2^j \cdot a$:

$$\left| \frac{1}{2^j t - k} \right|$$

has on $I = [-a, a]$ it's maximum at the point $t = a$.

So we get

$$\max_{t\in I}|\phi_{j,k}(t)| \leq \frac{2^{j/2}}{\pi}\frac{1}{k-2^j a}$$

and

$$|c_k| \leq \frac{2^{j/2}c}{\pi}\frac{1}{k-2^j a} .$$

Now we get:

$$\left\|\sum_k c_k\phi_{j,k} - \sum_{|k|\leq k_{max}} c_k\phi_{j,k}\right\| \leq \frac{2^{1+j/2}c}{\pi}\sqrt{\sum_{k=k_{max}+1}^{\infty}\frac{1}{(k-2^j a)^2}}$$

If we set $k_0 = k_{max} - 2^j a$ if a is an integer or generally $k_0 = k_{max} - [2^j a]$, then we get the proposition:

$$\left\|\sum_k c_k\phi_{j,k} - \sum_{|k|\leq k_{max}} c_k\phi_{j,k}\right\| \leq \frac{2^{1+j/2}c}{\pi}\sqrt{\sum_{k=k_0+1}^{\infty}\frac{1}{k^2}}$$

For a function with jumps at the edges of I (and if the jump is not very small) we get bad approximations not only with the Shannon wavelet, even with Daubechies wavelets, too. See [7].

For a approximation on a compact interval I we get much better results if we calculate a least square approximation on I or if we use a continuation of $1_I y$ which is continuous and quadratic integrable on R (see [6] and [7]).

References

[1] T.S. Carlson, J. Dockery, J. Lund (1997). *A Sinc-Collocation Method for Initial Value Problems*, Mathematics and Computation, Vol. 66, No. 217.

[2] Ricardo Estrada (1995). *Summability of cardinal series and of localized Fourier series.* Applicable Analysis: An International Journal

[3] J. R. Higgins (1985). *Five short Stories about the Cardinal Series.* American Mathematical Society

[4] Qian, L. (2002). *On the Regularized Whittaker-Koltel'nikov-Shannon Sampling Theorem.* Proceedings of the American Mathematical Society, Vol. 131, No. 4

[5] Schuchmann, M. (2012). *Approximation and Collocation with Wavelets. Approximations and Numerical Solving of ODEs, PDEs and IEs.* Osnabrück: DAV

[6] M. Schuchmann, M. Rasguljajew (2013). *An Approximation on a Compact Interval Calculated with a Wavelet Collocation Method can Lead to Much Better Results than other Methods.* Journal of Approximation Theory and Applied Mathematics (Vol. 1)

[7] M. Schuchmann, M. Rasguljajew (2013). *Approximation of Non $L^2(R)$ Functions on a Compact Interval with a Wavelet Base* (2013, Vol. 2)

[8] M. Schuchmann, M. Rasguljajew; (2014). *Fourier Properties of Approximations of Functions on a Compact Interval with Daubechies Wavelets.* International Journal of Emerging Technology and Advanced Engineering, Vol. 4, Issue 4, April 2014. (http://www.ijetae.com/files/Volume4Issue4/IJETAE_0414_46.pdf)

[9] J. M. Whittaker (1927). *On the Cardinal Function of Interpolation Theory.* Proceedings of the Edinburgh Mathematical Society.

Solving Fredholm integral equations with application of the four Chebyshev polynomials

Mostefa NADIR
Department of Mathematics University of Msila 28000 ALGERIA

Abstract

In this work, we study the approximation of the Fredholm integral equation of the second kind using the four Chebyshev series expansions. Those equations will be solved using m collocation points. That is to say, we will make the residual equal to zero at m points, giving us a system of m linear equations.

Key words Chebyshev polynomials, Fredholm integral equation, Collocation method, Numerical method.

2000 Mathematics Subject Classification: 45D05, 45E05, 45L05.

1. Introduction

Some phenomena which appear in many areas of scientific fields such as plasma physics, fluid dynamics, mathematical biology and chemical kinetics can be modelled by Fredholm integral equations [6]. Also this type of equations occur of scattering and radiation of surface water wave, where we can transform any ordinary differential equation of the second order with boundary conditions into a Fredholm integral equation.

$$\varphi(t_0) - \int_{-1}^{1} k(t,t_0)\varphi(t)dt = f(t_0), \tag{1}$$

with a given kernel $k(t,t_0)$ and a function $f(t)$, we try to find the unknown function $\varphi(t)$ as in [3,7] where the authors estimate the density function $\varphi(t)$ by means of Legendre and the first Chebyshev polynomials. For this study we replace the function $\varphi(t)$ by the four Chebyshev polynomials and compare the accuracy of the estimation of the unknown function with many numerical examples.

2. Discretization of integral equation

In this section, we apply a collocation method to the equation(1) in order to discredit and convert it to a system of linear equations. For this latter by using a Chebyshev polynomials we approximate the unknown $\varphi(t)$ such that

$$\varphi(t) = \varphi(t) = \frac{c_0}{2} + \sum_{n=1}^{\infty} c_n S_n(x), \tag{2}$$

where $S_n(x)$ denotes the nth Chebyshev polynomial of the first, second, third or fourth kind. So, that these series behave like Fourier series. Thus in particular, this series converges pointwise to φ on $[-1,1]$ if φ is continuous there, while the convergence is uniform if φ satisfies a Dini-Lipschitz condition or is of bounded variation, see, e.g. [1,2]. Then truncations of the series provide polynomials with good approximation properties on the interval $[-1,1]$..

The four Chebyshev polynomials with the interval of orthogonality $[-1,1]$, see [2,4] are defined as

1- **The first-kind polynomial T_n**

$$T_n(x) = \cos n\theta \quad \text{when} \quad x = \cos\theta$$

The three term recurrence formula satisfied by Chebyshev polynomials is the translation of the elementary trigonometric identity

$$\cos n\theta + \cos(n-2)\theta = 2\cos\theta \cos(n-1)\theta,$$

which becomes

$$T_n(x) = 2xT_{n-1}(x) - T_{n-2}(x), \quad n = 2, 3, ...$$

With

$$T_0(x) = 1, \quad T_1(x) = x$$

2- **The second-kind polynomial U_n**

$$U_n(x) = \frac{\sin(n+1)\theta}{\sin\theta} \quad \text{when} \quad x = \cos\theta$$

The three term recurrence formula satisfied by Chebyshev polynomials is the translation of the elementary trigonometric identity

$$\sin(n+1)\theta + \sin(n-1)\theta = 2\cos\theta \sin n\theta,$$

which becomes

$$U_n(x) = 2xU_{n-1}(x) - U_{n-2}(x), \quad n = 2, 3, ...$$

With

$$U_0(x) = 1, \quad U_1(x) = 2x$$

3- **The third-kind polynomial U_n**

$$V_n(x) = \frac{\cos(n+\frac{1}{2})\theta}{\cos\frac{1}{2}\theta} \quad \text{when} \quad x = \cos\theta$$

The three term recurrence formula satisfied by Chebyshev polynomials is the translation of the elementary trigonometric identity

$$\cos(n+\frac{1}{2})\theta + \cos(n-2+\frac{1}{2})\theta = 2\cos\theta \cos(n-1+\frac{1}{2})\theta,$$

which becomes

$$V_n(x) = 2xV_{n-1}(x) - V_{n-2}(x), \quad n = 2, 3, ...$$

With
$$V_0(x) = 1, \quad V_1(x) = 2x - 1$$

4- The fourth-kind polynomial W_n

$$W_n(x) = \frac{\sin(n+\frac{1}{2})\theta}{\sin\frac{1}{2}\theta} \quad \text{when} \quad x = \cos\theta$$

The three term recurrence formula satisfied by Chebyshev polynomials is the translation of the elementary trigonometric identity

$$\sin(n+\frac{1}{2})\theta + \sin(n-2+\frac{1}{2})\theta = 2\cos\theta \sin(n-1+\frac{1}{2})\theta,$$

which becomes

$$W_n(x) = 2xW_{n-1}(x) - W_{n-2}(x), \quad n = 2, 3, \ldots.$$

With
$$V_0(x) = 1, \quad V_1(x) = 2x + 1.$$

By substituting the relation (2) in the equation(1) we get

$$\sum_{n=0}^{m} c_n S_n(t_0) - \int_{-1}^{1} k(t, t_0) \sum_{n=0}^{m} c_n S_n(t_0) = f(t_0). \tag{3}$$

Choosing the equidistant collocation points as follows

$$t_j = -1 + \frac{2j}{m}, \quad j = 0, 1, \ldots m, \tag{4}$$

and define the residual as

$$R_n(t_0) = \sum_{n=0}^{m} c_n S_n(t_0) - \int_{-1}^{1} k(t, t_0) \sum_{n=0}^{m} c_n S_n(t) - f(t_0)$$

Then, by imposing conditions at collocation points

$$R_n(t_j) = 0, \quad j = 0, 2, \ldots m,$$

the integral equation (3) is converted to a system of linear equations.

Theorem
Let $A : X \to X$ be compact and the equation

$$(I - A)\varphi = f, \tag{5}$$

admit a unique solution. Assume that the projections $P_n : X \to X_n$ satisfy to $\|P_n A - A\| \to 0$, $n \to \infty$. Then, for sufficiently large n, the approximate equation

$$\varphi_n - P_n A \varphi_n * P_n f, \tag{6}$$

has a unique solution for all $f \in X$ and there holds an error estimate

$$\|\varphi - \varphi_n\| \leq M \|\varphi - P_n\varphi\|, \qquad (7)$$

with some positive constant M depending on A.

Proof

As it is known for all sufficiently large n the inverse operators $(I - P_nA)^{-1}$ exist and are uniformly bounded, see [1,5]. To verify the error bound, we apply the projection operator P_n to the equation (5) and get

$$P_n\varphi - P_nA\varphi = P_nf,$$

or again

$$\varphi - P_nA\varphi = P_nf + \varphi - P_n\varphi.$$

Subtracting this from (6) we find

$$(I - P_nA)(\varphi - \varphi_n) = (I - P_n)\varphi.$$

Hence the estimate (7) follows.

3. Numerical examples

Example 1

Consider the Fredholm integral equation

$$\varphi(t_0) - \int_{-1}^{1} \exp(2t_0 - \frac{5}{3}t)\varphi(t)dt = \exp(2t_0)(1 - 3\exp(\frac{1}{3}) + 3\exp(-\frac{1}{3}),$$

where the function $f(t_0)$ is chosen so that the solution $\varphi(t)$ is given by

$$\varphi(t) = \exp(2t)$$

The approximate solution $\widetilde{\varphi}(t)$ of $\varphi(t)$ is obtained by the solution of the system of linear equations for $N = 10$

Points of t	Exact solution	Approx solution	Error
-1.0000	1.353353e-001	1.353353e-001	2.915452e-008
-0.8000	2.018965e-001	2.018965e-001	4.349344e-008
-0.6000	3.011942e-001	3.011941e-001	6.488458e-008
-0.4000	4.493290e-001	4.493289e-001	9.679642e-008
-0.2000	6.703200e-001	6.703199e-001	1.444033e-007
0.0000	1.000000e+000	9.999998e-001	2.154244e-007
0.2000	1.491825e+000	1.491824e+000	3.213754e-007
0.4000	2.225541e+000	2.225540e+000	4.794358e-007
0.8000	3.320117e+000	3.320116e+000	7.152342e-007
0.8000	4.953032e+000	4.953031e+000	1.067004e-006
1.0000	7.389056e+000	7.389055e+000	1.591783e-006

Table 1. The exact and approximate solutions of example 1 in some arbitrary points, of the system of linear equations

Example 2
Consider the Fredholm integral equation

$$\varphi(t_0) - \int_0^1 (t_0 - t)^3 \varphi(t)dt = \frac{1}{1+2t_0^2} - \frac{t_0}{2}(6 + \sqrt{2}(-3 + 2t_0^2)\arctan(\sqrt{2})),$$

where the function $f(t_0)$ is chosen so that the solution $\varphi(t)$ is given by

$$\varphi(t) = \frac{1}{1+2t^2}.$$

The approximate solution $\widetilde{\varphi}(t)$ of $\varphi(t)$ is obtained by the solution of the system of linear equations for $N = 10$

Points of t	Exact solution	Approx solution	Error
-1.0000	3.333333e-001	3.333334e-001	2.597979e-008
-0.8000	4.385965e-001	4.385966e-001	1.046103e-007
-0.6000	5.813953e-001	5.813955e-001	1.251738e-007
-0.4000	7.575758e-001	7.575759e-001	1.043146e-007
-0.2000	9.259259e-001	9.259260e-001	5.867663e-008
0.0000	1.000000e+000	1.000000e+000	4.904144e-009
0.2000	9.259259e-001	9.259259e-001	4.035875e-008
0.4000	7.575758e-001	7.575757e-001	6.046791e-008
0.8000	5.813953e-001	5.813953e-001	3.877922e-008
0.8000	4.385965e-001	4.385965e-001	4.135147e-008
1.0000	3.333333e-001	3.333335e-001	1.965683e-007

Table 2. The exact and approximate solutions of example 2
in some arbitrary points, of the system of linear equations

4. Conclusion

In this work, a projection method known as collocation method with the four Chebyshev polynomials was chosen to discretize the Fredholm integral equations. This method has some advantages, there is no difference between the four Chebyshev polynomials. It is easy to require best convergence and less computations than other methods discussed in [3, 7]. In some methods the kernels of equations and the second member are required to satisfy some conditions.

5. References

[1] **K.E. Atkinson,** The numerical solution of integral equation of the second kind, Cambridge University press, (1997).

[2] **A. Benoit, B. Salvy,** Chebyshev expansions for solutions of linear differential equations, published in ISSAC'09 (2009)

[3] **K. Maleknejad, K. Nouri, M. Yousefi,** Discussion on convergence of Legendre polynomial for numerical solution of integral equations, Applied Mathematics and Computation 193 (2007) 335 339.

[4] **J. C. Mason, D.C. Handscomb,** Chebyshev polynomilas by CRC Press LLC 2003.

[5] **L. Kantorovitch, G. Akilov,** Functional analysis, Pergamon Press, University of Michigan (1982).

[6] **M. Nadir**, **A. Rahmoune**, Solving linear Fredholm integral equations of the second kind using Newton divided difference interpolation

polynomial,

[7] **L. Yucheng,** Application of the Chebyshev polynomial in solving Fredholm integral equations in Mathematical and Computer Modelling 50

(2009) 465 469

Fourier Properties of Approximations with Functions on a Compact Interval using Daubechies Wavelets

M. Schuchmann and M. Rasguljajew from the Darmstadt University of Applied Sciences

Abstract

One way to approximate a non $L^2(R)$ function g on a compact interval I is to use $y = 1_I \cdot g$ (with the indicator function 1_I). Here we calculate the orthogonal projection from $1_I \cdot g$ to V_j. With the factor 1_I we cut the function g and hence we generally get jumps on the edges of the interval I. This leads to a bad decay behavior of the base coefficients. In this paper we show that the orthogonal projection from $1_I \cdot g$ to V_j with Daubechies wavelet leads to a Fourier series and with a small j we get a too fast decay behavior of the coefficients of the approximation function y_j in comparison to y, which leads to all well known problems with Fourier approximations when we do not consider higher frequencies (e.g. oscillation effects).

Introduction

In the wavelet theory a scaling function ϕ is used, which belongs to a MSA (multi scale analysis). From the MSA we know, that we can construct an orthonormal basis of a closed subspace V_j, where V_j belongs to a the sequence of subspaces with the following property:

$$... \subset V_{-1} \subset V_0 \subset V_1 \subset ... \subset L^2(R),$$

$\{\phi_{j,k}(t)\}_{k \in Z}$ is an orthonormal basis of V_j with $\phi_{j,k}(t) = 2^{j/2} \phi(2^j t - k)$.

If we want to get an approximation of a function g on a compact interval I then g does not need be in $L^2(R)$ but must be in $L^2(I)$ if we use the function $1_I \cdot g$. But this type of function can lead to bad approximations what we will see soon. Generally it would be better to use a continuous or even better continuously differentiable extension of $1_I \cdot g$ on R, which also is quadratic integrable on R (see [6]).

We get an approximation function (as a orthogonal projection form y on V_j)

$$y_j(t) := \sum_k c_k \cdot \phi_{j,k}(t)$$

with

$$c_k = \langle y, \phi_{j,k} \rangle = \int_{-\infty}^{\infty} y(t) \cdot \overline{\phi_{j,k}(t)} dt \;.$$

Example:
We use the Daubechies wavelet of order $m = 4$ and calculate the orthogonal projection of $y(t) = e^{-t} \cdot 1_{[0,1]}(t)$ on V_2. For many functions without a jump we would get with $j = 2$ a very good approximation, but not here. Because of the compact support of the scaling function and the function y we only need for y_j a finite summation area for $k = -6, -5, ..., 3$.

Here we see thee graph of y with its approximation y_j:

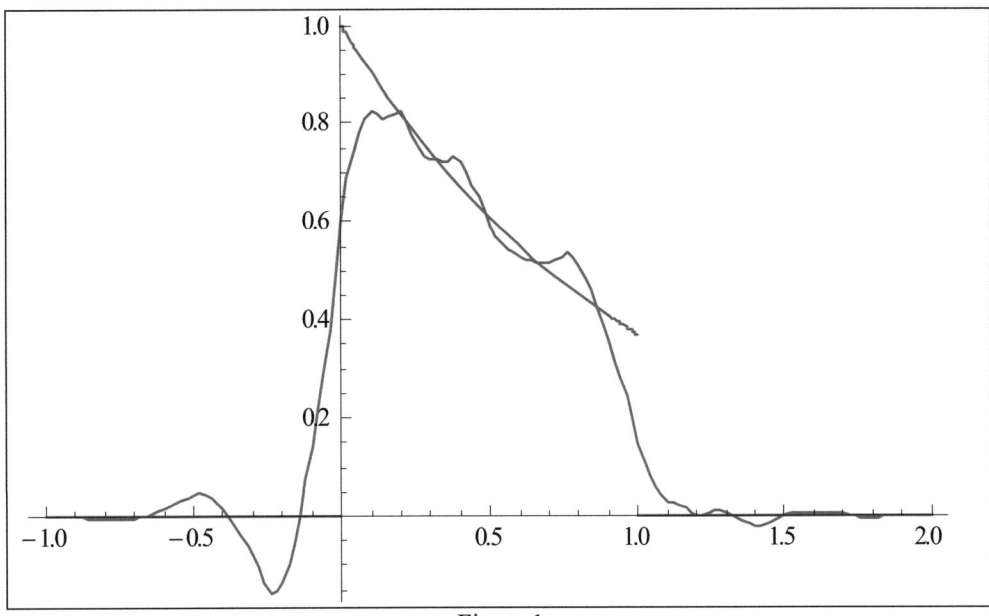

Figure 1

In [6] we described the same effect if we use the Shannon wavelet and we showed that the approximation function y_j by using the Shannon wavelet is almost a partial sum of a Fourier series with coefficients that have a bad decay behaviour. The Scaling function of the Shannon wavelet has a Fourier transform which is a rectangle function. The amplitude spectrum of the Fourier transform of the scaling function of the Daubechies wavelet of order 4 looks similar as a rectangle function (like all scaling functions of the Daubechies wavelet if the order m is not too small). It has no compact support but the decay is relative fast. That's one suggestion, but we well soon see, that even the best approximation of y in V_j is a Fourier series which has coefficients with a too fast decay behaviour in comparison to the Fourier series of the function y, so that we get a bad approximation if j is not very big.

Here is the amplitude spectrum of $\Phi^D_{2,k}$:

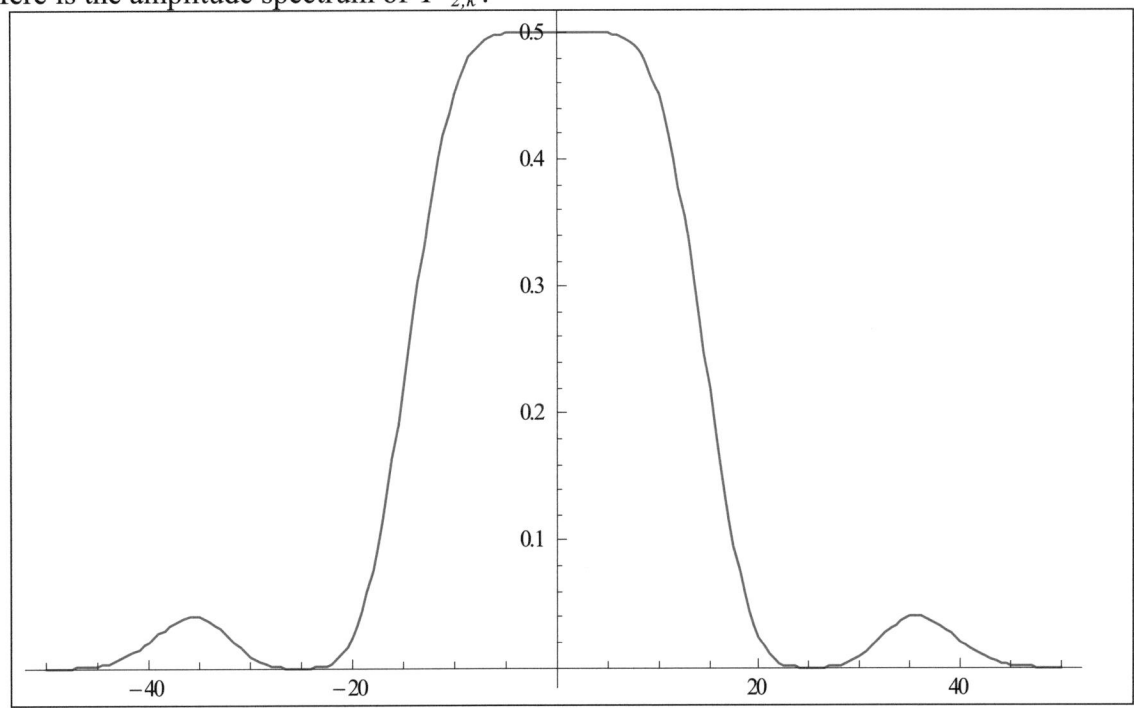

Figure 2

y and y_j and their Fourier Series

For the approximation of $y = 1_I \cdot g$ (with $I = [a, b]$) with a Daubechies wavelet of order m only a finite number of the coefficients c_k are unequal to zero because the scaling function $\phi^D{}_{j,k}$ of that Daubechies wavelet has compact support and $y = 1_I \cdot g$ too:

$$c_k = \langle f, \phi^D{}_{j,k} \rangle = \int_{-\infty}^{\infty} 1_I \cdot y(t) \cdot \overline{\phi^D{}_{j,k}(t)} dt = \int_a^b g(t) \cdot \overline{\phi^D{}_{j,k}(t)} dt$$

$$supp\ \phi^D = [0, 2m-1] \Rightarrow supp\ \phi^D{}_{j,k} = [2^{-j}k, 2^{-j}(2m-1+k)]$$

$c_k = 0$ for all k with $[2^{-j}k, 2^{-j}(2m-1+k)] \cap [a, b] = \{\}$ (or general only $\{x\}$ with a real x).

We assume that the biggest k with $c_k \neq 0$ is k_{max} and the smallest k is k_{min}.

So the best approximation or orthogonal projection of y on V_j is:

$$y_j(t) := \sum_{k=k_{min}}^{k_{max}} c_k \cdot \phi^D{}_{j,k}(t)$$

The base elements of V_j which are not equal to zero in $[a, b]$ are in a subspace of V_j:

That subspace is $span\{\phi^D{}_{j,k}\}_{k \in M}$ with $M = \{k_{min}, k_{min}+1, \ldots, k_{max}\}$ (or generally its closed hull). If we choose a smallest integer $\tilde{j}$, so that

$$\bigcup_{k \in M} supp\ \phi^D{}_{j,k} \subseteq [-2^{\tilde{j}} \cdot \pi, 2^{\tilde{j}} \cdot \pi]$$

then in the Fourier space all the $\{\phi^D{}_{j,k}\}_{k \in M}$ can be written as a Shannon series (with the base elements $\phi^S{}_{j,\tilde{k}}$ from the scaling function of the Shannon wavelet) and in the original space the Shannon series are Fourier series, so

$$\phi_{j,k}(t) = \frac{2^{-\tilde{j}/2}}{2\pi} \cdot 1_{[-2^{\tilde{j}} \cdot \pi, 2^{\tilde{j}} \cdot \pi]}(t) \cdot \sum_{\tilde{k}} b^k_{\tilde{k}} \cdot e^{-i \cdot t \cdot \tilde{k}/2^{\tilde{j}}} \qquad (1)$$

for almost all t (compare with [6]). If Φ^D is the Fourier transform of ϕ^D and $\Phi^D{}_{j,k}$ is the Fourier transform of $\phi^D{}_{j,k}$, we get the coefficients $b^k_{\tilde{k}}$ through

$$b^k_{\tilde{k}} = 2^{-\tilde{j}/2} \cdot \Phi^D{}_{j,k}(2^{-\tilde{j}} \cdot \tilde{k})$$

and therefore through:

$$b_{\tilde{k}}^{k} = \underbrace{2^{-\tilde{j}/2+j/2} \cdot \Phi^D(2^{-(j+\tilde{j})} \cdot \tilde{k})}_{=:B(\tilde{k})} \cdot e^{i \cdot 2^{-(\tilde{j}+j)} \cdot \tilde{k} \cdot k} \qquad (2)$$

Analogous we get the representation of *y*:

$$y(t) = \frac{2^{-\tilde{j}/2}}{2\pi} \cdot 1_{[-2^{\tilde{j}} \cdot \pi, 2^{\tilde{j}} \cdot \pi]}(t) \cdot \sum_r a_r \cdot e^{-i \cdot t \cdot r/2^{\tilde{j}}} \qquad (3)$$

with $a_r = 2^{\tilde{j}/2} \cdot Y(2^{-\tilde{j}} \cdot r)$.

So *y* can be written as a Fourier series but the coefficients a_r have the order $O(1/|r|)$ if *y* has jumps at the edges of *I* as in the example above.

So we would get for the coefficients c_k:

$$c_k = \frac{1}{2\pi} \sum_r a_r \cdot \overline{b_r^k} \qquad (4)$$

For easier notation we leave out the factor $1_{[-2^{\tilde{j}} \cdot \pi, 2^{\tilde{j}} \cdot \pi]}(t)$ and we consider that $t \in [-2^{\tilde{j}} \cdot \pi, 2^{\tilde{j}} \cdot \pi]$. No we get with (1):

$$y_j(t) = \sum_{k=k_{min}}^{k_{max}} c_k \cdot \phi^D{}_{j,k}(t)$$

$$= \frac{2^{-\tilde{j}/2}}{2\pi} \cdot \sum_{k=k_{min}}^{k_{max}} c_k \cdot \sum_{\tilde{k}} b_{\tilde{k}}^k \cdot e^{-i \cdot t \cdot \tilde{k}/2^{\tilde{j}}}$$

$$= \frac{2^{-\tilde{j}/2}}{2\pi} \cdot \sum_{\tilde{k}} e^{-i \cdot t \cdot \tilde{k}/2^{\tilde{j}}} \cdot \underbrace{\sum_{k=k_{min}}^{k_{max}} c_k \cdot b_{\tilde{k}}^k}_{:=a_{\tilde{k}}^{approx}}$$

$$= \frac{2^{-\tilde{j}/2}}{2\pi} \cdot \sum_{\tilde{k}} a_{\tilde{k}}^{approx} \cdot e^{-i \cdot t \cdot \tilde{k}/2^{\tilde{j}}}$$

So we see that y_j can be written as a Fourier series, but in *y* the coefficients $a_{\tilde{k}}^{approx}$ have in comparison to the coefficients $a_{\tilde{k}}$ a too fast decay behaviour. So with $a_{\tilde{k}}^{approx}$ we do not consider the 'higher frequencies'. That is what we see if we take a look at $a_{\tilde{k}}^{approx}$. With (2) we get:

$$a_{\tilde{k}}^{approx} = \sum_{k=k_{min}}^{k_{max}} c_k \cdot b_{\tilde{k}}^k = B(\tilde{k}) \cdot \sum_{k \in M} c_k \cdot e^{i \cdot 2^{-(\tilde{j}+j)} \cdot \tilde{k} \cdot k} \qquad (5)$$

$B(\widetilde{k})$ is calculated with the function values $\Phi^D(2^{-(j+\widetilde{j})}\cdot\widetilde{k})$. Φ^D is the Fourier transform of the scaling function of order m from the Daubechies wavelet. Φ^D has (with a not too small m) a very strong decay behaviour and for a small j the factor $B(\widetilde{k})$ does too. So the needed higher frequencies are not considered.

Using (4), (5) and (2) we can write the coefficients $a_{\widetilde{k}}^{approx}$ with a_r:

$$a_{\widetilde{k}}^{approx} = \frac{1}{2\pi}B(\widetilde{k})\cdot\sum_r a_r \cdot \overline{B(r)} \cdot \sum_{k\in M} c_k \cdot e^{i\cdot 2^{-(\widetilde{j}+j)}\cdot(\widetilde{k}-r)\cdot k}$$

Conclusion

We saw, that we can write the approximation function from the Daubechies wavelets as a fourier series, if the function has compact support. If the function has jumps like generally functions of the type $1_I \cdot g$, then y_j is for small j a bad approximation because the decay of the fourier coefficients is too fast and so we get oscillation effects.

References

[1] Ricardo Estrada (1995). *Summability of cardinal series and of localized Fourier series.* Applicable Analysis: An International Journal

[2] J. R. Higgins (1985). *Five short Stories about the Cardinal Series.* American Mathematical Society

[3] Qian, L. (2002). *On the Regularized Whittaker-Koltel'nikov-Shannon Sampling Theorem.* Proceedings of the American Mathematical Society, Vol. 131, No. 4

[4] Schuchmann, M. (2012). *Approximation and Collocation with Wavelets. Approximations and Numerical Solving of ODEs, PDEs and IEs.* Osnabrück: DAV

[5] M. Schuchmann, M. Rasguljajew (2013). *An Approximation on a Compact Interval Calculated with a Wavelet Collocation Method can Lead to Much Better Results than other Methods.* Journal of Approximation Theory and Applied Mathematics (Vol. 1)

[6] M. Schuchmann, M. Rasguljajew (2013). *Approximation of Non $L^2(R)$ Functions on a Compact Interval with a Wavelet Base* (2013, Vol. 2)

[7] M. Schuchmann, M. Rasguljajew; (2013). *Error Estimations in an Approximation on a Compact Interval with a Wavelet Bases.* COMPUSOFT - An international journal of advanced computer technology, Vol. 2, Issue 11, November 2013.

[8] J. M. Whittaker (1927). *On the Cardinal Function of Interpolation Theory.* Proceedings of the Edinburgh Mathematical Society.